YOUR KNOWLEDGE HAS VALUE

- We will publish your bachelor's and master's thesis, essays and papers

- Your own eBook and book - sold worldwide in all relevant shops

- Earn money with each sale

Upload your text at www.GRIN.com
and publish for free

Bibliographic information published by the German National Library:

The German National Library lists this publication in the National Bibliography; detailed bibliographic data are available on the Internet at http://dnb.dnb.de .

This book is copyright material and must not be copied, reproduced, transferred, distributed, leased, licensed or publicly performed or used in any way except as specifically permitted in writing by the publishers, as allowed under the terms and conditions under which it was purchased or as strictly permitted by applicable copyright law. Any unauthorized distribution or use of this text may be a direct infringement of the author s and publisher s rights and those responsible may be liable in law accordingly.

Imprint:

Copyright © 2018 GRIN Verlag
Print and binding: Books on Demand GmbH, Norderstedt Germany
ISBN: 9783668695597

This book at GRIN:

https://www.grin.com/document/424140

Sultan Alshammari

Toxicological effects of carbon nanotubes in animals

GRIN Verlag

GRIN - Your knowledge has value

Since its foundation in 1998, GRIN has specialized in publishing academic texts by students, college teachers and other academics as e-book and printed book. The website www.grin.com is an ideal platform for presenting term papers, final papers, scientific essays, dissertations and specialist books.

Visit us on the internet:

http://www.grin.com/

http://www.facebook.com/grincom

http://www.twitter.com/grin_com

Toxicological effects of carbon nanotubes in animals

Sultan Alshammari

Outlines:

1. **Introduction**
   - History of carbon nanotubes,
   - Discovery of carbon nanotubes
   - Results of carbon nanotubes and effects in animals

2. **Effects of carbon nanotubes**
   - Cardiovascular effects
   - Effects on the liver and the spleen
   - Effects on the airways
   - Irritation and the nervous system
   - Skin and Reproduction
   - Genotoxic effects and Carcinogenicity
   - Short-term and long-term effects after repeated exposure (up to 90 days)
   - Summary and discussion

Introduction

Carbon nanotubes are a revolutionary discovery, and new carbon structure and have recently had a surprising instant. Carbons nanotubes (CNT's) were first discovered back in 1983 by Japanese physicists Sumio Lijima since their discovery many scientists have worked together to unravel the secrets behinds his incredible carbon structure that seems never to stop surprising us with their versatility. Among many exciting properties of carbon nanotubes, one of the most useful ones is the fact that they can take a role of the metal or a semiconductor. Metals are known for their electrical and thermal conductivity, most of metals behavior are well known but semiconductors are a bit more complicated to explain. One amazing property about CNT is they are incredibly fast collectivity where nowadays people can direct them truthfully using metals wire that service rose for life used to travel with carbon nanotubes. We can grade wires that can change speed several times faster than conventional wires. Certain kind of nanotubes act as electron highways they form baths for electric to travel and CNT are incredibly high conductivity. Scientists have confirmed CNT are officially stronger than steel and that makes CNT the best materials for building nano structures that are sure to its that any possible application. It is not hard to believe why these structure are as strong as they are. Because of strong bonds between these carbons, CNTs are stronger than steel, 270 – 950 Gpa modulus of carbon nanotubes where 200 Gpa of steel. However, CNT's toxicity have been investigated over last decay since many industries start develop CNT's and apply them on their products. Human and animals are exposers to CNT's each day where its so questionable materials.

Carbon nanotubes have different structures which different properties between them. Single-walled-Carbon nanotube (SWCNT) have unidimensional structure, and it's high surface area which is useful for drug-loading capacity. SWCNTs are efficient for drug delivery than Multi-Walled-Carbon nanotubes (MWCNT), both of them have different toxicity based on their structures, and each structure has his property. For instance, (SWCNT) when its production needs metal as a catalyst like nickel and results in high metal content lead to worker or environment expose too. Moreover, we must include techniques that carbon nanotubes are produced for determining what might affect the CNT's. First, Chemical vapor deposition is common and wide manufacture way of production carbon nanotubes, where due to low cost, high yield, purity and easy to scale up.

However, this technique is based on thermals decomposition of gas hydrocarbon with metals as a catalyst, like nickel, cobalt, yttrium, and iron. With gases contains carbons like carbon monoxide and methane get into an oven and heat up, the heat up is the reaction of the heated gas with metals as catalysts. Both MWCNT and SWCNT are producing form this technique and MWCNT have the better quality where the only problem with this technique is a large amount of pollutant up to 40%, and these consist of amorphous carbon, nano-structure graphite and carbon encapsulated metals particles from the catalysts (Köhler et al., 2008). Another manufacturing technique is Arc discharge; this was used with the discovery of carbon nanotubes (Iijima,1991) this technique is based on the application of high voltage field over two graphite rods, anode and cathode are applied to arc discharge for stability. The carbon nanotubes are grown with cathode where the anode is being consumed. The distant is constant between cathode and anode, but anode is adjustable, the process takes place in helium gas. The production of SWCNT to be obtained the electrodes must be doped with the small amount of metals particles that act as catalysts. Overall, this method is cheap and used to generate carbon nanotubes with high quality but high impurity as well. A third technique, laser ablation is a production of carbon nanotube that introduces by Smalley and his co-workers (Guo et al., 1995a), this method is first used to generate MWCNT but with refined. For instance, catalytic particles like cobalt and nickel can be used for the production of SWCNT. Laser ablation graphite is kept at 1200 in a chamber, where laser pulsed with high energy used to vaporize the graphite carbons. Carbon nanotubes are growing on cooled in the chamber which slowly flows into the chamber. If the graphite is pure MWCNT is generated where to generate SWCNT need to dope Cobalt or Nickel. The toxicity of carbon nanotubes is exist based on tuns of studies, where some found the toxicity of pollutants that come with carbon nanotubes is because of the manufacturing process (Shvedova et al., 2003).

The effect of carbon nanotubes on cell and animals experiments where it's difficult to differentiate the result from studies, which the minimum lethal dose of carbon nanotubes is known for animals, another aspect is carbon nanotubes physical or chemical properties are not adequately described in most studies. Diameter, length, surface area/unit weight and degree of purity are a characterization of determining if a type of carbon nanotubes is toxic or not. How animals exposed to carbon nanotubes is also essential, where different experimental exposure result indifferent.

Methods have an advantage and disadvantages in animals like Installation, and inhalation methods are different in results, and the reason for that is the methods of exposures (Li et al., 2007a). The mortality of experimental animals is being published when 0.5 mg of SWCNT through respiratory passage caused death 5 of 9 animals (Lam et al., 2004). Where 1,000 mg of carbon nanotubes did not cause death in all mice that treated which above doses according to the writer of the articles (Kolosnjaj-Tabi et al., 2010). Animals experiments have shown that carbon nanotubes may affect the heart and blood vessels after doped in respiratory passages, in this concept the effect of carbon nanotubes are being listed below:

Cardiovascular effects:

The possibility of carbon nanotubes to affect heart and blood vessels is exist and that via increase of stress related-genes and this gene involved in the recruitment of immune cell and increasing markers for oxidative stress. Then, it grew in length aorta, and cardiac tissue along with an impact in regulating blood pressure in the arteries and that will result in reducing blood flow to coronary arteries and decline the heart tissue. The effect on cardio-vascular most was not because nanotubes transferring through passage but were indirect effect for which exact mechanism is unknown, where it is suggested that the increase of production of cytokines decreased persuade by CNT's can be the reason for systematic effect. Table 1 shows several studies in Cardiovascular effects after exposure to MWCNT, and SWCNT.

Table 1: Cardiovascular effects after exposure to MWCNT, and SWCNT.

Absolute dose	Dose (mg/kg)	Animal	Number	Diameter	Length	Characteristics	Exposure form	Exposure time	Effects	Reference	Type
-	0,064	Rat	8-10	0,9-1,7 nm	<1 μm	-	Tube feed	1 dose, 24 hrs	Modified nucleotides in the DNA of the lung and liver	Folkmann et al. 2009	SWCNT
10 μg	0,3	Mouse	-	-	-	Unfunctionalized	Intratracheal	1 dose	No significant effects	Tong et al. 2009	SWCNT
10 μg	0,4	Mouse	-	0,7-1,5 nm	1 μm	-	Pharyngeal instillation	1 dose, 60 days	Damage to mitochondrial DNA in the aorta	Li et al. 2007b	SWCNT
-	0,64	Rat	8-10	0,9-1,7 nm	<1 μm	-	Tube feed	1 dose, 24hrs	Modified nucleotides in the DNA of the lung and liver	Folkmann et al. 2009	SWCNT
-	1	Rat	8	1,2-1,6 nm	2-5 nm (sic!)	-	Intratracheal	2 does over 4 weeks	Modified baroreflex, the influence of autonomic cardiovascular control	Legramante et al. 2009	SWCNT
40 μg	1,3	Mouse	-	-	-	Unfunctionalized	Intratracheal	1 dose	Increase of creatine kinase, cardiac fibre degeneration	Tong et al. 2009	SWCNT
40 μg	1,3	Mouse	-	-	-	Acid-functionalized	Intratracheal	1 dose	Significant increase in size of infarction, creatine kinase, cardiac fibre degeneration	Tong et al. 2009	SWCNT
10 μg	1,3	Mouse	-	-	-	Acid-functionalized	Intratracheal	1 dose	No significant effects	Tong et al. 2009	SWCNT
40 μg	1,4	Mouse	-	0,7-1,5 nm	1 μm	-	Pharyngeal instillation	1 dose, 60 days	Damage to mitochondrial DNA	Li et al. 2007b	SWCNT
40 μg	1,5	Mouse	-	0,8-1,2 nm	0,1-1 μm	8,8 wt% Iron	Pharyngeal aspiration	4 h	Local and systemic effect on inflammatory parameters	Ederly et al. 2009	SWCNT
40 μg	1,5	Mouse	-	80 nm	10-20 μm	0,2% wt% Iron	Pharyngeal aspiration	4 h	Local and systemic effect on inflammatory parameters	Ederly et al. 2009	MWCNT
20 μg x 4	2,8	Mouse	-	0,7-1,5 nm	1 μm	-	Pharyngeal instillation	4 does over 8 weeks	Atherosclerosis was increased in ApoE-/- transgenic mice. Plaque formation in aorta, mtDNA damage	Li et al. 2007b	SWCNT

Effects on the liver and the spleen:

By the existence of carbon nanotubes CNT's in the blood, it will reach liver and spleen. The noxious effect of non-functionalized MWCNT in the liver leads to increase the level of inflammatory cells and injury markers with the death of liver tissue (Ji et al., 2009, Zhang et al., 2010). The effect is based on the concentration of doses which refer to low doses is not affect the liver or not toxic. Inflammation and fibrosis of the liver are observed in mice that treat with 250 mg/kg MWCNT in the stomach cavity. The amount 250 mg is high compare that with other studies like MWCNT has been functionalized with PEG (Polyethylene glycol) resulted significantly less impact indicate that the toxicity rest on what the surface of CNT's looks like come behind oral exposure. Oxidative DNA damage was found in rat liver after injecting carbon nanotube with that much of weight (Folkmann et al., 2009).

Effects on the airways:

The effect of CNT's on airways has been studied throughout instillation and inhalation in laboratory animals. After inhalation of MWCNT, an increase in inflammatory cells is being resulted as with increases levels of injury markers. Another study MWCNT could pass through the lung, the pleura, when treated with 30 mg for six hours exposure. Meanwhile, pleural fibrosis has been noticing which mean that the connective tissue in the outer portion is being covered. Comparable observation has been made after MWCNT is introduced in airways of mice, carbon nanotubes were found in pleura. The penetration of the lung barrier depends on the dose of carbon nanotubes (Mercer et al., 2010). Fibrosis and CNT's that located over the pleura were also observe after the treatment of MWCNT in other studies (Porter et al., 2010). There are studies states that granulomas of carbon nanotubes treatment have been reported on airways of experimental animals. One study compares the effect of instillation and inhalation of SWCNT and MWCNT, conclude that inhalation has the worse effect than installation (Shvedova et al., 2008b). In other studies, researchers have demonstrated that MWCNT and SWCNT can affect the effectiveness of the cells, that means the permeability barrier of the epithelial cells might be harmed or harmful that could leave a penetration between cells and reach to lungs tissue. Long carbon nanotubes and short do not affect 5-9um MWCNT and 0.5 um SWCNT (Rotoli et al., 2009; Rotoli et al., 2008). Table 1 and 2 shows Effects of airway exposure to SWCNT and MWCNT.

Table1 Effects: of airway exposure to SWCNT

Absolute dose	Dose (mg/kg)	Animal	Number	Diameter	Length	Characteristics	Exposure form	Exposure time	Effects	Reference
1 µg	0.005	Rat	6	20-50 nm	0.5-2 µm	Dispersed with BSA	Instillation	1 dose	Significant increase in phagocytosis 1 and 7 days after exposure	Elgrabli et al. 2008
	0.04	Rat	-	60 nm	1.5 µm	Unmodified	Instillation	1 dose	No effects after 6 months	Kobayashi et al. 2010
10 µg	0.05	Rat	6	20-50 nm	0.5-2 µm	Dispersed with BSA	Instillation	1 dose	Significant increase in phagocytosis 1-90 days after exposure	Elgrabli et al. 2008
0.3 mg/m³	0.2	Mouse	6	10-20 nm	5-15 µm	Unmodified	Inhalation	6 h/day	Repeated inhalation 7-14 days. Carbon nanotubes in the lung macrophages, no increase in white blood cells. Systemic immunosuppression	Mitchell et al. 2007
	0.2	Rat	-	60 nm	1.5 µm	Unmodified	Instillation	1 dose	Observations 1 day - 6 months after dosing. No effects	Kobayashi et al. 2010
1 mg/m³	0.2	Mouse	10	10-50 nm	1-10 µm	-	Inhalation	En 6 h exposure	Observed 1-14 days. No effects	Ryman-Rasmussen et al. 2009a
0.1 mg/m³ (46.8 µg/animal)	0.23-0.3	Rat	10	5-15 nm	0.1-10 µm	9.6% AlO	Inhalation	6h/day on 65 occasions for 90 days	Mild granulomatous inflammation. No systemic effects. (LOEC)	Ma-Hock et al. 2009
1 mg/m³	0.5	Mouse	6	10-20 nm	5-15 µm	Unmodified	Inhalation	6 h/day	Repeated inhalation 7-14 days. Carbon nanotubes in the lung macrophages, no increase in white blood cells. Systemic immunosuppression	Mitchell et al. 2007
10 µg	0.3	Mouse	6-8	49 nm	3.9 µm	Unmodified	Aspiration	1 dose	Observed 1-56 days. CNTs, which penetrated the alveolar epithelium	Mercer et al. 2010
100 µg	0.5	Rat	6	20-50 nm	0.5-2 µm	Dispersed with BSA	Instillation	1 dose	Increase in inflammatory cells after 7 and 180 days, significant phagocytosis 1-180 days	Elgrabli et al. 2008
20 µg	1	Mouse	12	0.8-1.2 nm	100-1000 nm	Unmodified, 17.7% Fe-content	Instillation	1 dose	Increased number of inflammatory cells, increase of inflammatory proteins	Shvedova et al. 2008b
40 µg	1.3	Mouse	4	-	Aggregates approx 25-150 nm	Treated with acid or not	Aspiration	1 dose	Increased number of inflammatory cells, increase of inflammatory proteins	Saxena et al. 2007
40 µg	1.6	Mouse	4-8	1-2 nm	100-2000 nm	Unmodified, dissolved in PBS	Instillation	1 dose	After 30 days, inflammation and granuloma	Mutlu et al. 2010
40 µg	1.6	Mouse	4-8	1-2 nm	100-2000 nm	Unmodified, dissolved in Pluronics	Instillation	1 dose	Well separated CNTs caused no effects. Excretion via macrophages	Mutlu et al. 2010
40 µg	2	Mouse	6	1-4 nm	1-3 µm	Unmodified	Aspiration	1 dose + bacterial 10 days	Increase in inflammatory cells, synergistic increase in collagen after CNT bacterial exposure, impaired phagocytosis of bacteria	Shvedova et al. 2008a

Table2: Effects of airway exposure to MWCNT

Absolute dose	Dose (mg/kg)	Animal	Number	Diameter	Length	Characteristics	Exposure form	Exposure time	Effects	Reference
40 µg	2	Mouse	6	1-4 nm	1-3 µm	Unmodified	Aspiration	1 dose/10 days	Increase in inflammatory cells, increase in collagen	Shvedova et al. 2008a
0,1 mg	3,3	Mouse	5	-	-	High levels of metal contaminants	Instillation	1 dose	Slight inflammation and granulomas in animals treated with contaminated CNT	Lam et al. 2004
-	4	Mouse	8	2-20 nm	100 nm - several µm	Unmodified	Instillation	1 dose	Treatment with CNT and 33 µg/kg LPS led to worsening pneumonia. Proinflammatory proteins increased in the blood.	Inoue et al. 2008
0,5 mg	16,5	Mouse	5	-	-	High levels of metal contaminants	Instillation	1 dose	More severe inflammation and granuloma, mortality 5/9 in a group	Lam et al. 2004
0,5 mg	-	Mouse	-	-	-	-	Instillation	1 dose	Activation of macrophages, the presence of granulomas, oxidative stress	Chou et al. 2008
50 µg × 7	1,6 × 7	Mouse	12-13	0,8-2 nm	100 nm -15 µm	-	Instillation	1 dose	Carbon nanotubes exacerbated allergic response in mice. T-helper chemo-and cytokines increased	Inoue et al. 2010
20 µg	0,9	Mouse	6-8	49 nm	3,9 µm	Unmodified	Aspiration	1 dose	Observed 1-56 days. Granulomatous ulcers, CNT, which penetrated the alveolar epithelium	Mercer et al. 2010
-	1	Rat	-	40-60 nm	0,5-500 µm	Unmodified	Instillation	1 dose	Mild inflammation	Liu et al. 2008

Absolute dose	Dose (mg/kg)	Animal	Number	Diameter	Length	Characteristics	Exposure form	Exposure time	Effects	Reference
-	1	Rat	-	60 nm	1,5 µm	Unmodified	Instillation	1 dose	Observations 1 day - 6 months after dosing. Transient inflammation. MWCNT in macrophages. Slight granuloma formation	Kobayashi et al. 2010
0,5 mg/m³ (243 µg/djur)	1,2-1,6	Rat	10	5-13 nm	0,1 -10 µm	9,6% AlO	Inhalation	6h/day on 65 occasions for 90 days	Granulomatous inflammation, inflammation, lipoproteinosis. No systemic effects	Ma-Hock et al. 2009
0,05 mg	1,6	Mouse	5-6	50 nm	10 µm	Unmodified	Instillation	1 dose	Inflammation of the bronchi, damage to the alveoli, immune response	Li et al. 2007a
40 µg	1,8	Mouse	7-8	49 nm	3,9 µm	Unmodified	Aspiration	1 dose	Observed 1-56 days. Granulomatous ulcers CNT which penetrated the aveolar epithelium	Mercer et al. 2010
0,5 mg	2,2	Rat	4-6	9,7 nm	5,9 µm	Unmodified	Instillation	1 dose	No pathological findings, 81% of the dose remaining after 60 days	Muller et al. 2005
0,5 mg	2,2	Rat	4-6	11,3 nm	0,7 µm	Ground	Instillation	1 dose	Increase of TNF-alpha and soluble type I collagen, 36% in a dose remaining after 60 days	Muller et al. 2005
0,07 mg	2,3	Mouse	5-6	50 nm	10 µm	Unmodified	Inhalation	-	Repeated inhalation for 8 days. Proliferation and increase in the thickness of the alveoli	Li et al. 2007a
5 mg/m³	2,7	Mouse	6	10-20 nm	5-15 µm	Unmodified	Inhalation	6 h/day	Repeated inhalation 7-14 days. Carbon nanotubes in the lung macrophages, no increase in white blood cells. Systemic	Mitchell et al. 2007

Irritation and the nervous system:

Increase the irritation of carbon nanotubes is not clear, wherein some eye irritation studies show that Draize test has been done on rabbit for 72 hours resulting in no irritation. Where other studies indicate irritation of the eyes were obtained. (Kishore et al., 2009). The effect of carbon nanotubes on the nervous system has been investigated over experimental animals by injecting carbon nanotubes into the brain, where the distribution of electrical activity is being observed. Another way is by tube feeding where ultrastructure studies showed the feeding carbon nanotubes mainly in lysosomes and some in mitochondria (Yang et al., 2010).

Skin and Reproduction:

Nowadays, carbon nanotubes are unknown whether it can penetrate via status corneum of the skin and reach further into the skin. Carbon nanotube was applied to the skin of experimental animals resulting in inflammation of the epidermis and dermis. The cause of damage is that an increase in reactivity of oxygen radicals, where it is the reason oxidative stress (Murray et al., 2009). Study have shown that carbon nanotubes influence cells death, inflammatory proteins, and oxidative stress. Resulted in contradiction to negatives effects are also available, where are no impacts found were found on laboratory animals or cultured cells (Huczko and Lange, 2001, Kishore et al., 2009). The effect of carbon nanotubes in reproduction is limited, where injection MWCNT of the water-soluble result in absorption in male mice. Carbon nanotubes could have an effect on zebrafish, where MWCNT cause defects at low concentration, and embryonic death, delayed hatching, cell death and abnormal spinal cord growth at high concentration (Asharani et al., 2008). Table 3 shows effects after skin exposure to MWCNT and SWCNT.

Table 3:. Effects after skin exposure to MWCNT and SWCNT.

Absolute dose	Dose (mg/kg)	Animal	Number	Diameter	Length	Characteristics	Exposure form	Exposure time	Effects	Reference	Type
0.5 mg	-	Rabbit	-	-	-	Dimension: 901 nm	Dermal	4 h	None	Kishore et al. 2009	MWCNT
0.5 mg	-	Rabbit	-	-	-	Dimension 554 nm	Dermal	4 h	None	Kishore et al. 2009	MWCNT
0.1 mg	0.6	Rat	-	20-50 nm	220 nm	Unmodified	Subcutaneous	1-4 weeks	Slight inflammation, no neutrophils, no necrosis	Sato et al. 2005	MWCNT
0.1 mg	0.6	Rat	-	20-50 nm	825 nm	Unmodified	Subcutaneous	1-4 weeks	Slight inflammation, no neutrophils, no necrosis	Sato et al. 2005	MWCNT
40 µg	2.35	Mouse	3	-	-	30% Iron	Dermal	1 dose/day, 5 days	No observed effects	Murray et al. 2009	SWCNT
80 µg	4.7	Mouse	3	-	-	30% Iron	Dermal	1 dose/day, 5 days	Several cells in the epidermis, oxidative damage	Murray et al. 2009	SWCNT
160 µg	9.4	Mouse	3	-	-	30% Iron	Dermal	1 dose/day, 5 days	More cells in the epidermis, thickening of the skin, collagen production, oxidative damage	Murray et al. 2009	SWCNT

Genotoxic effects and Carcinogenicity:

The genetic effect caused by carbon nanotubes have been studied on cultured cells. In many cases, the genetic impact have considered, and so called micronuclei, an effect of both MWCNT and SWCNT and the reason is ambiguous, and chromosome separation during cell division is affected by metal contamination in carbon nanotubes because nanotubes have the same dimension as cytoskeleton of the cell (Cveticanin et al., 2010; Zhu et al., 2007; Ochoa- Olmos et al., 2009; Migliore et al., 2010; Sargent et al., 2009; Patlolla et al., 2010; Lindberg et al., 2009; Jacobsen et al., 2008). Also, some studies show no genetoxicity of carbon nanotubes, and that might be because the nature of carbon nanotubes used, purity and separation of particles in solution are another aspects. The shaping of granuloma or mesothelioma after MWCNT in the abdominal cavity has been demonstrated in some studies where they suggested that carbon nanotubes could have the same asbestos effects by producing the same fiber effect, comes to contact with mesothelioma layers which line outside body organs. Based on one study, the formation of mesothelioma used mice were muting p53, which is tumor protein that controls cell division and develops cancer (Takagi et al., 2008). Along-term exposure of single dose MWCNT did not approve for developing cancer, but researcher points out that the study should interpret precisely as the absence of reaction does not mean that CNT's can not cause cancer. A possibility is nanotubes were to short to leave any carcinogen impact. Table 4 show Carcinogenic effects after exposure to MWCNT or SWCNT.

Table 4 Carcinogenic effects after exposure to MWCNT or SWCNT.

Absolute dose	Dose (mg/kg)	Animal	Number	Diameter	Length	Characteristics	Exposure form	Exposure time	Effects	Reference	Type
-	1	Rat	7	70-110 nm	2 μm	Unmodified	Intrascotal	1 dose, 1 year	Mesotheliom, 6/7 animals died	Sakamoto et al. 2009	MWCNT
0,5 mg	2-2,5	Rat	5	11,3 nm	0,7 μm	Ground	Intratracheal	1 dose, 3 days	Inflammation, neutrophils and macrophages	Muller et al. 2008	MWCNT
50 μg	2,7	Mouse	3-6	15 nm	1-5 μm	Short	Intraperitoneal	1 dose, 7 days	non significant inflammation, no granulomas	Poland et al. 2008	MWCNT
50 μg	2,7	Mouse	3-6	10,4 nm	5-20 μm	Short	Intraperitoneal	1 dose, 7 days	no significant inflammation, 1 granuloma in 1 animal	Poland et al. 2008	MWCNT
50 μg	2,7	Mouse	3-6	85 nm	13 μm	Long	Intraperitoneal	1 dose, 7 days	significant inflammation, granuloma in abdominal cavity mesotel	Poland et al. 2008	MWCNT
50 μg	2,7	Mouse	3-6	165 nm	56 μm	Long	Intraperitoneal	1 dose, 7 days	significant inflammation, granuloma in abdominal cavity mesotel	Poland et al. 2008	MWCNT
2 mg	5,6-7,1	Rat	50	11,3 nm	0,7 μm	Unmodified	Intraperitoneal	1 dose, 2 years	None	Muller et al. 2009	MWCNT
2 mg	5,6-7,1	Rat	50	11,3 nm	0,7 μm	Structural defects	Intraperitoneal	1 dose, 2 years	None	Muller et al. 2009	MWCNT
2 mg	8-10	Rat	5	11,3 nm	0,7 μm	Ground	Intratracheal	1 dose, 3 days	Inflammation, neutrophils and macrophages. Induction of micronuclei in pneumocyte	Muller et al. 2008	MWCNT
5 mg	20-25	Rat	5	11,3 nm	0,7 μm	Ground	Intratracheal	1 dose, 3 days	Inflammation, neutrophils and macrophages.	Muller et al. 2008	MWCNT
20 mg	55-71	Rat	50	11,3 nm	0,7 μm	Structural defects	Intraperitoneal	1 dose, 2 years	None	Muller et al. 2009	MWCNT
20 mg	55-71	Rat	50	11,3 nm	0,7 μm	Unmodified	Intraperitoneal	1 dose, 2 years	None	Muller et al. 2009	MWCNT
3 mg	111-121	Mouse	19	100 nm	5 μm	Unmodified	Intraperitoneal	1 dose, 180 days	Mesothelioma, 14/16 animals died	Takagi et al. 2008	MWCNT

Short-term and long-term effects after repeated exposure (up to 90 days):

There are few studies in long-term and how can carbon nanotubes affect in vivo with repeated exposure. Most of the researchers have been studied using single-dose exposure upon experimental animals were to differentiate the length of time, the longest so far is being for two years. Experimental animals have been studied for three months period, where inhalation of MWCNT at the doses range from 0.1 to 6 mg/m^3 for 6 hours a day, five days a week, 13 weeks in total. No impact was found at 0.1 to 6 mg/m^3, but there are inflammation and epithelial damage at 0.4 to 6 mg/m3 (Pauluhn, 2010). In another similar study, rat inhaled MWCNT over three months where its result in developing granulomas and exhibit inflammation in airways. The strongest effect was found to be at 0.5 to 2.5 mg/m3 and that because of the impact that effect has already been observed at 0.1 mg/m3 and it was not possible for determining no observable effect level. Low repeated dosed shows an effect and this is important for risk assessment of carbon nanotubes. Another study is when mice repeatedly exposed for 7-14 days by inhaling the similar dose of MWCNT where no effect was observed, also inhibition of systematic immune defense was found (Mitchell et al., 2007).

Summary and discussion:

In many studies, a correlation has been done between the biological effect and the dose of the carbon nanotubes, where Table1, summarizes the connection between the dose given to experimental animals or cell cultures and observed effects. Some of the investigation on animals and cells that not included in this report because it will be described in tables later on in this report. In numerous studies, respiratory being exposed to carbon nanotubes lead to the formation of grandmas, a transition of carbon nanotubes to the pleura, and fibrosis. This occurs at low concentration CNT in the air, 0.1 mg/m3 and enhancement of fibrosis occur at 0.25mg/kg body weight after multi-able exposures of SWCNT on mice. Increase the dosage lead to the worsening conditions(look to tables below). For skin exposures, the effect on experimental animals can be seen at 4.7mg/kg or 0.6mg/kg of the injection of SWCNT and MWCNT. There is doses-respond for surface dermal exposure to SWCNT. (Table 2) Carbon nanotubes so far have caused mesothelioma (cancer) of the pleura in experimental animals. Mesothelioma arises after the intraperitoneal injection in mice, and the scrotum of rats and the effect take place at 111-121 mg/kg body weight of doses in mice and 1 mg/kg in rats. Overall, The toxicity of carbon nanotubes is debatable topic or concept for the last two decays, where as illustrated in this report most stated that carbon nanotubes cause cancers and brain tumors. Where other impacts are inflammation of the skin, damage DNA, and increase inflammatory cells in airways. More importantly, tuns of publication are taken about safety and protection device, since this report is about the effects of carbon nanotubes, doesn't be included.

References:

Asharani PV, Serina NGB, Nurmawati MH, Wu YL, Gong Z & Valiyaveettil S (2008) Impact of Multi-Walled Carbon Nanotubes on Aquatic Species. Journal of Nanoscience and Nanotechnology 8:3603-3609

Huczko A & Lange H (2001) Carbon nanotubes: Experimental evidence for a null risk of skin irritation and allergy. Fullerene Science and Technology 9:247-250

Folkmann JK, Risom L, Jacobsen NR, Wallin H, Loft S & Moller P (2009) Oxidatively Damaged DNA in Rats Exposed by Oral Gavage to C-60 Fullerenes and Single-Walled Carbon Nanotubes. Environmental Health Perspectives 117:703-708

Guo T, Nikolaev P, Rinzler Ag, Tomanek D, Colbert Dt & Smalley Re (1995a) Self-Assembly of Tubular Fullerenes. Journal of Physical Chemistry 99:10694-10697

Guo T, Nikolaev P, Thess A, Colbert Dt & Smalley Re (1995b) Catalytic Growth of Single-Walled Nanotubes by Laser Vaporization. Chemical Physics Letters 243:49-54

Guo JX, Zhang X, Li QN & Li WX (2007) Biodistribution of functionalized multiwall carbon nanotubes in mice. Nuclear Medicine and Biology 34:579-58

Iijima, S. (1991). Helical microtubules of graphitic carbon. Nature, 354(6348), 56-58. doi:10.1038/354056a0

Ji ZF, Zhang DY, Li L, Shen XZ, Deng XY, Dong L, Wu MH & Liu YF (2009) The hepatotoxicity of multi-walled carbon nanotubes in mice. Nanotechnology 20

Köhler AR, Som C, Helland A, Gottschalk F (2008) Studying the potential release of carbon nanotubes throught the application life cycle. J Cleaner Production 16: 927-937

Kolosnjaj-Tabi J, Hartman KB, Boudjemaa S, Ananta JS, Morgant G, Szwarc H, Wilson LJ & Moussa F (2010) In Vivo Behavior of Large Doses of Ultrashort and Full-Length Single-Walled Carbon Nanotubes after Oral and Intraperitoneal Administration to Swiss Mice. Acs Nano 4:1481-1492

Lam CW, James JT, McCluskey R & Hunter RL (2004) Pulmonary toxicity of single-wall carbon nanotubes in mice 7 and 90 days after intratracheal instillation. Toxicological Sciences 77:126-134

Li JG, Li WX, Xu JY, Cai XQ, Liu RL, Li YJ, Zhao QF & Li QN (2007a) Comparative study of pathological lesions induced by multiwalled carbon nanotubes in lungs of mice by intratracheal instillation and inhalation. Environmental Toxicology 22:415-421

Mercer RR, Hubbs AF, Scabilloni JF, Wang LY, Battelli LA, Schwegler-Berry D, Castranova V & Porter DW (2010) Distribution and persistence of pleural penetrations by multi-walled carbon nanotubes. Particle and Fibre Toxicology 7

Mitchell LA, Gao J, Wal RV, Gigliotti A, Burchiel SW & McDonald JD (2007) Pulmonary and systemic immune response to inhaled multiwalled carbon nanotubes. Toxicological Sciences 100:203-214

Murray AR, Kisin E, Leonard SS, Young SH, Kommineni C, Kagan VE, Castranova V & Shvedova AA (2009) Oxidative stress and inflammatory response in dermal toxicity of single-walled carbon nanotubes. Toxicology 257:161-171

Rotoli BM, Bussolati O, Barilli A, Zanello PP, Bianchi MG, Magrini A, Pietroiusti A, Bergamaschi A & Bergamaschi E (2009) Airway barrier dysfunction induced by exposure to carbon nanotubes in vitro: which role for fiber length? Human & Experimental Toxicology 28:361-368

Rotoli BM, Bussolati O, Bianchi MG, Barilli A, Balasubramanian C, Bellucci S & Bergamaschi E (2008) Non-functionalized multi-walled carbon nanotubes alter the paracellular permeability of human airway epithelial cells. Toxicology Letters 178:95-102

Shvedova AA, Castranova V, Kisin ER, Schwegler-Berry D, Murray AR, Gandelsman VZ, Maynard A & Baron P (2003) Exposure to carbon nanotube material: Assessment of nanotube cytotoxicity using human keratinocyte cells. Journal of Toxicology and Environmental Health-Part A 66:1909- 1926

Shvedova AA, Fabisiak JP, Kisin ER, Murray AR, Roberts JR, Tyurina YY, Antonini JM, Feng WH, Kommineni C, Reynolds J, Barchowsky A, Castranova V & Kagan VE (2008a) Sequential exposure 70to carbon nanotubes and bacteria enhances pulmonary inflammation and infectivity. American Journal Of Respiratory Cell And Molecular Biology 38:579-590

Zhang DY, Deng XY, Ji ZF, Shen XZ, Dong L, Wu MH, Gu TY & Liu YF (2010) Long-term hepatotoxicity of polyethylene-glycol functionalized multi-walled carbon nanotubes in mice. Nanotechnology 21

Yang Z, Zhang YG, Yang YLA, Sun L, Han D, Li H & Wang C (2010) Pharmacological and toxicological target organelles and safe use of single-walled carbon nanotubes as drug carriers in treating Alzheimer disease. Nanomedicine-Nanotechnology Biology and Medicine 6:427-441

YOUR KNOWLEDGE HAS VALUE

- We will publish your bachelor's and master's thesis, essays and papers

- Your own eBook and book - sold worldwide in all relevant shops

- Earn money with each sale

Upload your text at www.GRIN.com and publish for free